José Daniel Acosta Robles
Mario Alberto De La Cruz Padilla
Luis Alberto Morales Alias

MECHANICAL VIBRATIONS

José Daniel Acosta Robles
Mario Alberto De La Cruz Padilla
Luis Alberto Morales Alias

MECHANICAL VIBRATIONS

Cases documented in plant

ScienciaScripts

Cover image: www.ingimage.com

This book is a translation from the original published under ISBN 978-620-2-24226-4.

Publisher:
Sciencia Scripts
is a trademark of
Dodo Books Indian Ocean Ltd. and OmniScriptum S.R.L publishing group

120 High Road, East Finchley, London, N2 9ED, United Kingdom
Str. Armeneasca 28/1, office 1, Chisinau MD-2012, Republic of Moldova, Europe
Printed at: see last page
ISBN: 978-620-6-12957-8

Contents

ACKNOWLEDGEMENTS

I would like to thank all the people who supported the realisation of this work: especially the technical staff who with their skill and experience carry out the work in each of the areas; the predictive maintenance department of Ingenio Industrial Azucarera San Cristobal S.A. de C.V. for providing the tools and sharing the technical knowledge of each of its members.

PROLOGUE

Condition-based maintenance currently has a major impact on productivity in most industries, especially because it allows the actual condition of an asset to be monitored in order to decide what maintenance should be carried out.

Mechanical vibration analysis is used to monitor trends and detect problems by means of spectrum diagnosis, without the need to stop production lines or critical equipment within the plant, measuring damage or breakdowns by measuring the amplitude of vibrations; all this in order to schedule maintenance shutdowns and organise: required spare parts, personnel and corrective activities, avoiding prolonged shutdowns due to unforeseen events that increase production costs.

Within the sugar industry, lost time means an important delay in sugar production, since sugar cane as a raw material is scheduled for a certain time (zafra), and production costs increase with every hour of delay.

Each of the cases is presented in order to study the most common problems based on the topical issues for the diagnosis of vibration spectra analysis.

It is for this reason that this manual arises, with the aim of helping to diagnose the state of the equipment taking as a comparative example real cases and documented events using vibration analysis and being complemented with other predictive maintenance techniques such as:

- Visual inspection (strobe lamp).
- Infrared thermography.
- Alignment of shafts by laser method.
- Rotor balancing.
- Oil analysis.
- Mounting of bearings by induction heating.
- Demodulation (bearing envelope).

For this work we concentrated information, documents, courses, literature from authors and suppliers to facilitate the understanding of each of the methods used in each situation for a solution in a technical and accurate way, with the collaboration of the experience of the technical staff working in each of the Mechanical, Electrical and Machining areas.

CASE 1

UNBALANCE IN
HORIZONTAL SWING-TYPE BLADE-HOLDING ROTOR
BACK
(UNBALANCE)

The analysed system consists of a horizontal rotor of approximately 120 inches long, which carries 6 bars on which are mounted 22 swing back blades (which are not fixed) on each bar, with an approximate weight of 20 kilograms each blade, making a total of 132 blades, the weight of the assembled rotor is approximately 8 tons, the drive equipment consists of a 1,000 HP turbine operating at a speed of 3,900 CPM and a speed reducer with a ratio of 4.77, the rated speed of the rotor blade set is 817 CPM under load.

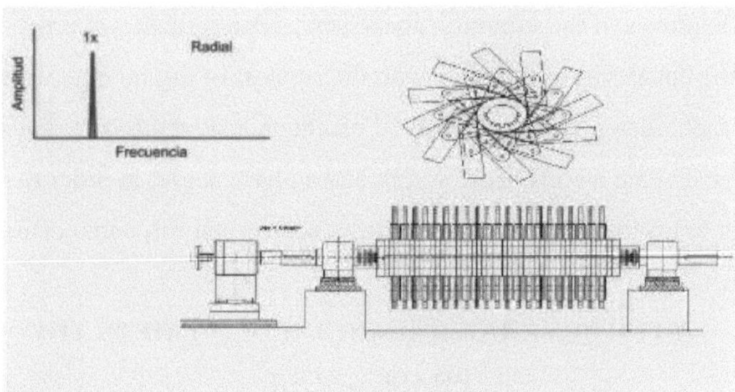

Fig. 1.1.- Horizontal blade-holding rotor and its vibration amplitude.

Field data: A high amplitude vibration value was perceived at 817 CPM (1x) in horizontal direction (0.8 in/s), values out of range, it is worth mentioning that blades have been recovered, re-weighed and welded against weights to ensure they maintain a uniform weight, these counter weights sometimes fall off due to the impact of the high loads entering the conductor causing unbalance.

In the vibration spectrum, high amplitude could be observed at the rotor rotation frequency of 817 CPM (1x), followed by harmonics at (2x), (3x) and (4x) in the presence of mechanical clearances, the highest value taken on the free side bearing (0.63 in/s). The amplitudes due to mechanical backlash were three times lower than the rotational speed.

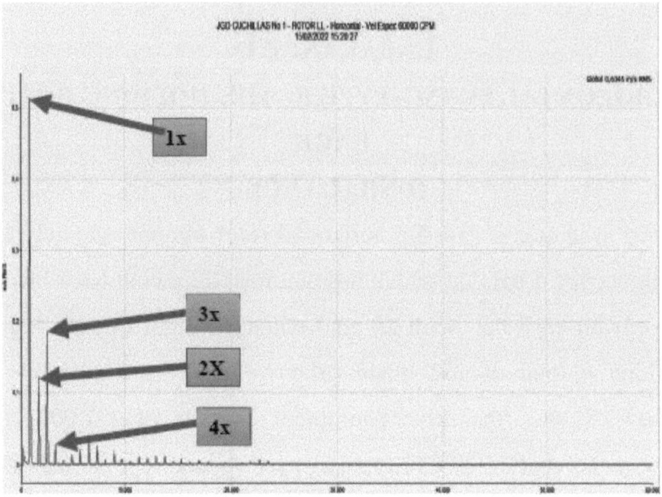

Fig. 1.2.- Vibration spectrum of the horizontal rotor.

Once the diagnosis of the vibration analysis was determined, we proceeded to carry out the balancing of the rotor with the support of digital equipment and a balancing tachometer, which consists of placing a test weight at a random angle, in order to calculate a correction weight and a phase angle, in order to find a weight that helps to counteract the resulting centrifugal force that causes imbalance.

AFTER SEVERAL BALANCING ATTEMPTS THE PHASE ANGLE WAS ALWAYS UNSTABLE AND THE AMPLITUDE RESPONSE DID NOT DECREASE.

It is recommended for this case before the assembly of the chopper blades, to check and ensure that the weight of these is uniform or have a maximum deviation of 5 grams, it is also important to verify that the size of the blades is uniform, this is done so that the main axis of inertia of the rotor coincides with the geometric axis of the system. It is also important to improve the anchoring of the bearings in order to eliminate structural clearances.

Follow-up: After several days of operation, a complete and new blade change was requested, with blades of the same weight and dimensions, retightening of the anchors and base plate, revision of the adjustment of the fixing sleeves and lubrication of the equipment. Previously, considerable weight differences could be observed, as well as different sizes of blades before the change affecting the balancing and operation of the equipment.

An unbalance in such a heavy rotor will, if allowed to operate for a prolonged period of time, result in system damage such as: loosening of anchorages, loosening of base plates, loosening of locking sleeves, breakage of welds and rigid equipment, high G's, damage to couplings and overheating of bearings.

After the complete replacement of the blades, it was observed that the rotation frequency value of 817 CPM (1x) which predominates in the spectrum was lower in amplitude than before (0.6 in/s), dropping considerably (0.14 in/s) although it is within the operating parameters, it could be improved by checking the alignment of the shafts between the rotor and the output shaft of the gearbox, as well as by improving the anchoring of the bearing bases.

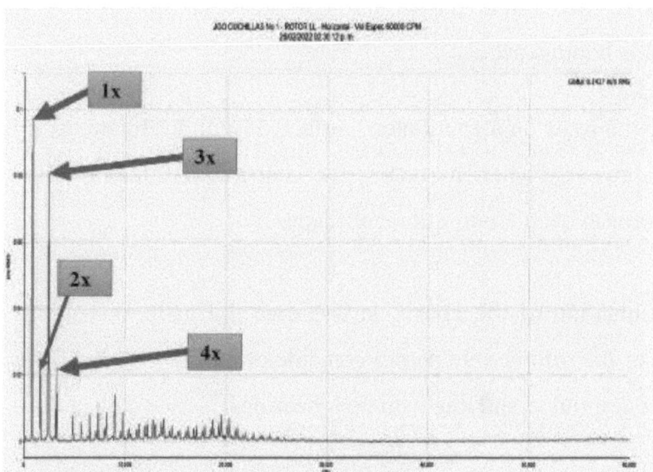

Fig. 1.3.- Rotor vibration spectrum after blade change.

FOLLOW-UP: DEFINITIONS REGARDING THE PHENOMENON OF UNBALANCE OR IMBALANCE IN ROTORS

Mechanical imbalance

Mechanical unbalance is the most common source of vibration in systems with rotating elements; hence, we can say that every rotor maintains a level of residual unbalance. These generate vibrations or not, depending on whether they work within the quality tolerances established in the rotor speed and characteristics standards.

CAUSES OF MECHANICAL IMBALANCE

Many factors can cause mechanical unbalance in a rotating machine. However, we can present some of the most common ones:

1) Replacement of rotor components during maintenance operations.
2) Deposits of matter accumulated in the operation of the machine.
3) Rotor distortion due to temperature.
4) Uneven wear of materials.

TYPES OF IMBALANCE

Static

It is when the main axis of a mass is displaced parallel to the geometrical axis of the rotor. The main axis passes through the centre of gravity if the mass distribution is homogeneous.

Pair of forces

If we place the rotor on a frictionless surface, it will not rotate, as it is statically balanced. However, when it rotates, the centrifugal forces of the two points cause the rotor to oscillate in a conical shape.

Dynamic

It is a combination of the above, in this case the main axis of mass and the centre line of the rolling axle do not coincide or intersect. It has different values of vibration amplitude and angle on both bearings.

CASE 2

STEAM TURBINE ROTOR UNBALANCE

(IMBALANCE AND MECHANICAL LOOSENESS)

The analysed system consists of a two-stage steam turbine, with a power of 1,400 HP and a working speed of 3,850 CPM, which transmits through a 1035G20 coupling to a collinear shaft speed reducer, with a power of 1,500 HP and a speed ratio of 4.795:1 coupled on the output shaft to a blade rotor with an approximate weight of 8 tons.

Field data: A high amplitude vibration value (0.41 in/s) was observed in the horizontal direction at a rotational frequency of 3,850 CPM (1x) coming from the turbine rotor, physically in the structure vibration with a sensation of wobble is perceived, no other frequency is observed in the spectrum.

It is worth mentioning that the load that the equipment receives is irregular and depends on the power supply that the driver operator handles.

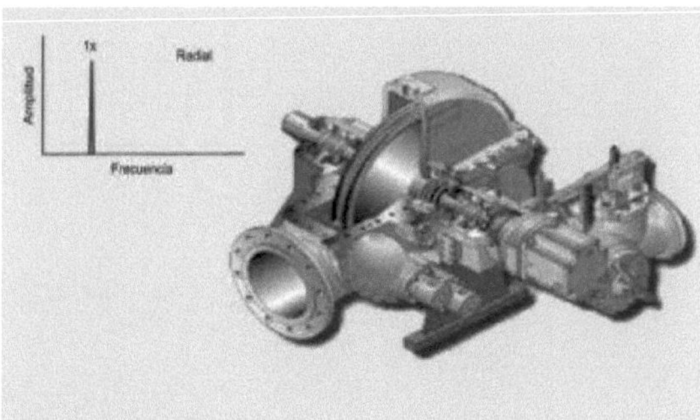

Fig. 2.1.- Two-stage steam turbine.

The vibration spectrum shows an amplitude value (0.44 in/s) in horizontal direction at the speed of rotation of the turbine rotor (1x), an imbalance was perceived, the possible causes could be: loosening of the anchor, loosening of screws or imbalance in the rotor due to incrustations of contaminants from the steam or a blow to some blade of the turbine rotor, it is recommended that each of the possible causes be ruled out and the rotor be checked at the appropriate time.

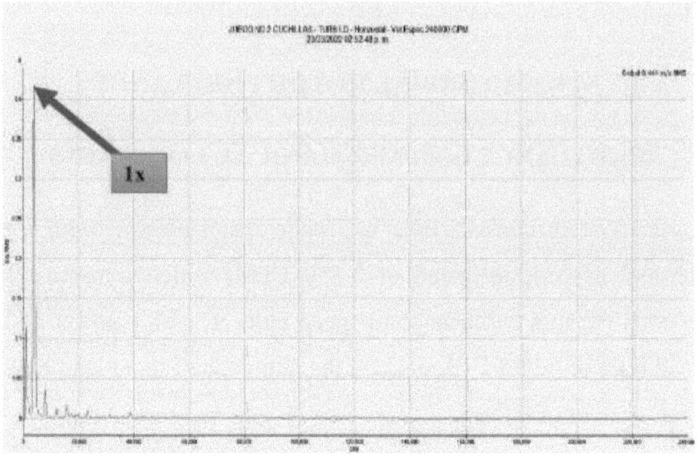

Fig. 2.2.- Vibration spectrum of the two-stage steam turbine.

If a low amplitude value (0.18 in/s) is observed in the vibration spectrum in the vertical direction, check the axial values to rule out any misalignment, when the unbalance appears in high values in the radial direction (horizontal and vertical) the axial values are usually very low.

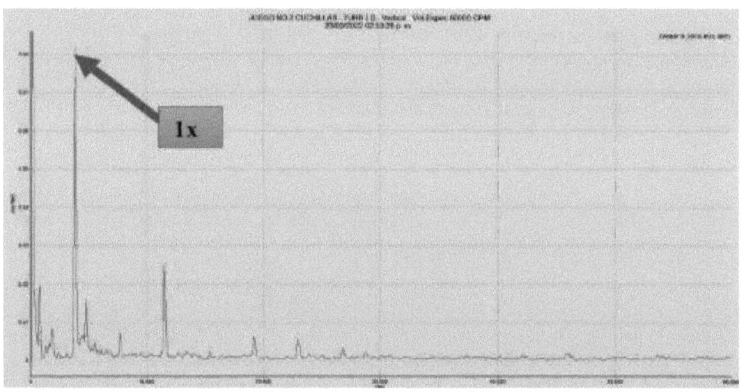

Fig. 2.3.- Vibration spectrum with low amplitude value (0.18 in/s) in vertical direction.

The unbalance evaluated from the waveform should be closely sinusoidal, otherwise there may also be misalignment, slack or other incorrect condition in addition to the unbalance. Phase is the best indication, look for a 90° phase shift between vertical and horizontal. The phase at the ends of the rotor bearings will be between 30° and 150° out of phase.

With the support of the vibration equipment, a waveform is observed, as a defined sine wave can be seen.

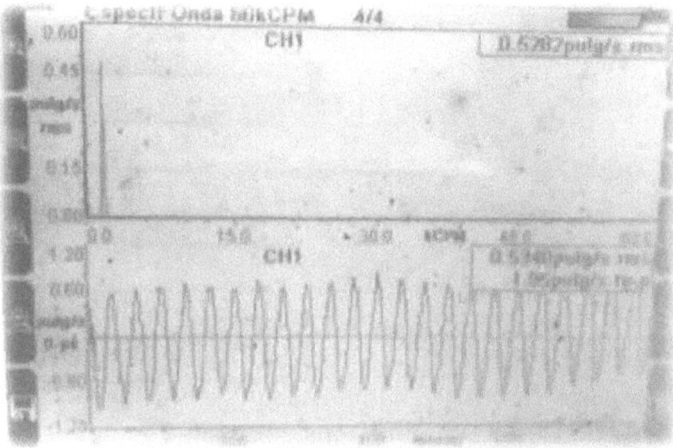

Fig. 2.4.- Vibration spectrum obtained with the analyser.

WAVEFORM, BEHAVIOUR OF THE VIBRATION WITH RESPECT TO TIME.

The unbalance is corrected with a mass compensation at the right point, with a vibration analyser you can locate the weight and angle to place or remove this compensation mass.

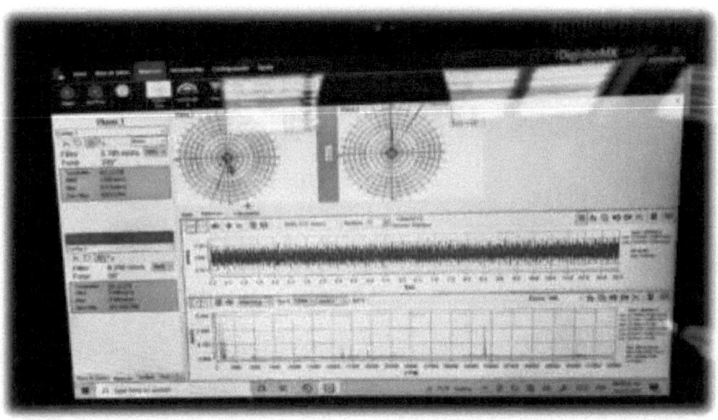

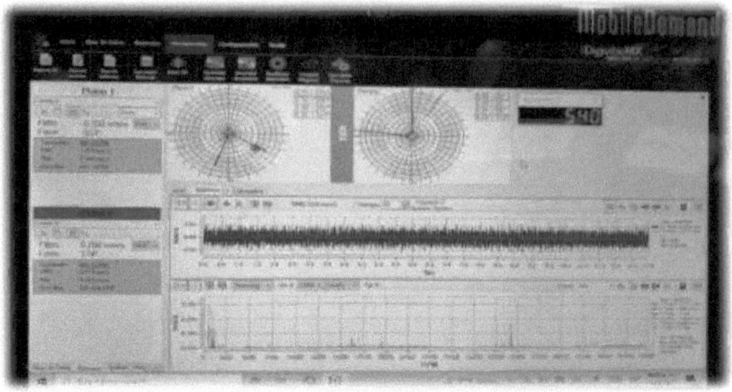

Fig. 2.5.- Vibration spectrum obtained with the analyser.

During a scheduled shutdown to change a steam supply valve, the steam turbine was uncovered.

Fig. 2.6.- Steam turbine rotor.

The rotor settings and condition of the rotor were checked, and for greater certainty the turbine rotor was sent to an external workshop for a diagnosis of the rotor balancing.

Fig. 2.7.- Steam turbine rotor balancing method.

We took advantage of the scheduled shutdown to check the tightness of the base plate anchorages, as well as the turbine and reducer fixing bolts, finding the beginning of looseness in the speed reducer bolts, this could be a cause derived from the high vibration caused by the rotor. Mass that was removed to balance the rotor.

In addition to the balancing of the turbine rotor, the shaft alignment between the turbine rotor and the speed reducer inlet pinion is verified by taking tolerances for the 1035G20 coupling.

A thermal growth of the turbine shaft of 0.005" was considered, it is recommended to leave 0.005" below the centre of the pinion shaft of the speed reducer, so that in normal operation (275 °C steam temperature) the centres of each shaft are aligned.

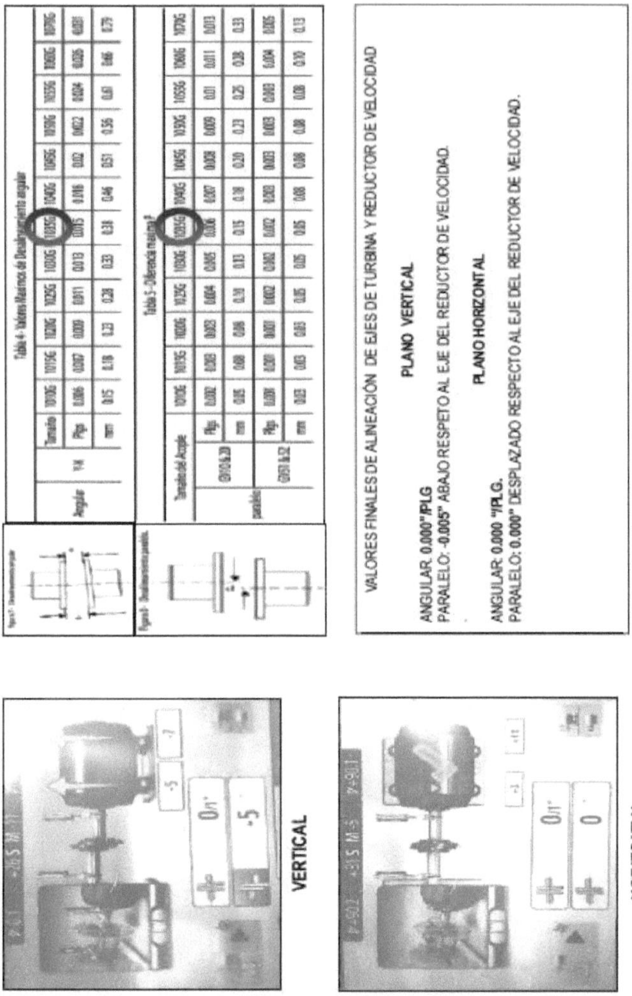

Fig. 2.8.- Final values of turbine shaft alignment y speed reducer

The vibration spectrum is shown after the verification and correction works shown above, it is possible to appreciate the decrease of the global value of the amplitude (from 0.41 in/s to 0.19 in/s) in the horizontal direction, acceptable amplitude values for the correct operation of the equipment.

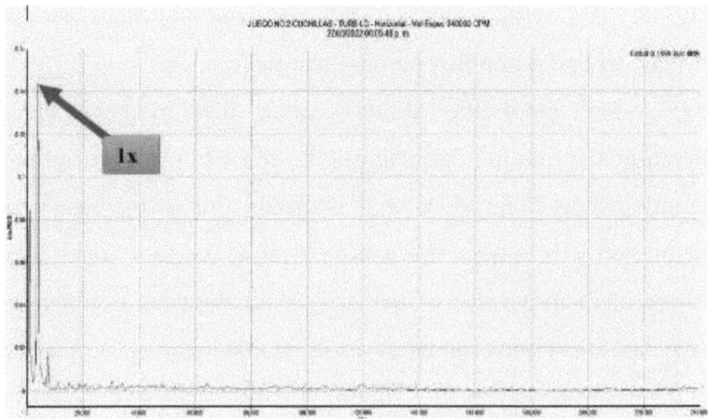

Fig. 2.9.- Vibration spectrum of the steam turbine after repairs.

CASE 3

SOFT FOOT OR LAME FOOT IN ENGINE ELECTRICO (TYPE "A" MECHANICAL BACKLASH)

The system analysed consists of a 150 HP induction motor with an asmcronic speed of 1,780 CPM, with a flexible 11 F grid coupling that drives a 6x8-15, 1,800 GPM centrifugal pump that pumps clear juice.

Field data: A high amplitude vibration value (0.94 in/s) was perceived in vertical direction at a rotation frequency of 1,780 CPM (1x), the tightness of the engine mountings was checked, as well as the fixation of the base plate, out of route the vibration values were taken in vertical in the four legs of the engine, finding in one of them a value of 0.4 in/s, concluding the presence of a lame foot, movement of the shims can be observed in that leg.

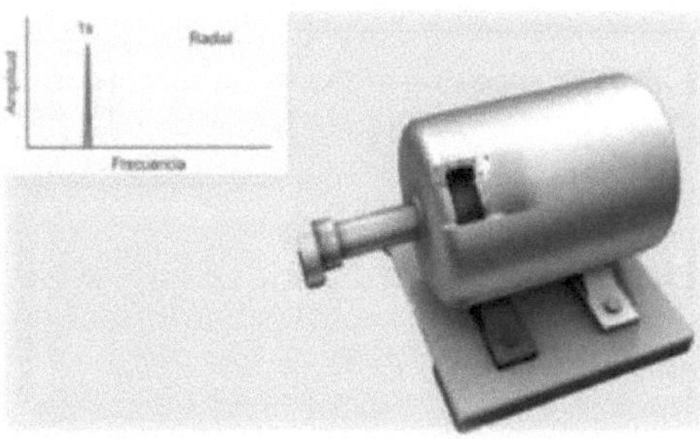

Fig. 3.1.- 150 HP Indian motor for clear juice pumping.

Spectrum: Structural weakness will generate a strong (1x) in the direction of the greatest weakness, unless there is an impact, in which case there will also be harmonics. To distinguish looseness, unbalance or resonance: if the horizontal (1x) is more than twice the amplitude of the vertical (1x), looseness is suspected. If the amplitude ratio changes when the speed of the machine is changed, then resonance should be suspected.

Phase: Structural loosening will often have a vibrating component (the motor foot) and a stationary component (the motor base); therefore, look for a 180° phase difference between the stationary and vibrating component.

Waveform: The waveform will be fairly sinusoidal (in speed) if there is any contact, the top and/or bottom of the wave may be truncated.

Soft or lame footing occurs in machines when the feet of the machine and the platform on which it is mounted are not on the same plane.

In electric motors, a lame foot distorts the frame, which in turn can distort the stator magnetic field. This creates unbalanced electrical forces between the rotor and the stator magnetic field.

Lame foot is best checked with a caratula indicator to determine the amount of lame foot and corrected with feeler gauges. Lame foot on motors can lead to premature bearing failure or loose or broken rotor bars.

Vibration spectrum with high amplitude value (0.94 in/s) at the rotational speed of 1,780 CPM (1x), in vertical direction, caused by the instability of the electric motor (soft foot) on one of the motor legs, causing the motor to behave like a pendulum up and down in vertical direction, is shown.

It is recommended to correct the alignment starting with the elimination of the soft or lame foot of the engine, this is done by shimming with shims trying to fill or compensate for the space created by deformation of the seat of one of the legs of the engine.

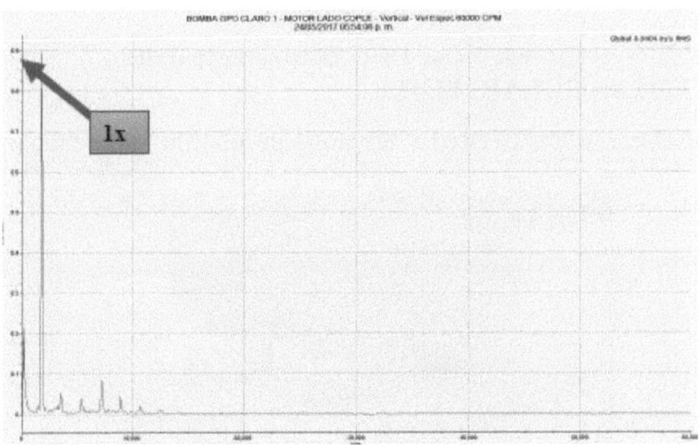

Fig. 3.2.- Vibration spectrum (0.94 in/s), rotational speed of 1,780CPM (1x).

It shows spectrum after correction of soft or lame foot on one of the electric motor legs, as well as verification of the correct tightening of the four motor legs to their base, achieving a considerable decrease of the vibration in the vertical direction from 0.94 in/s to 0.28 in/s.

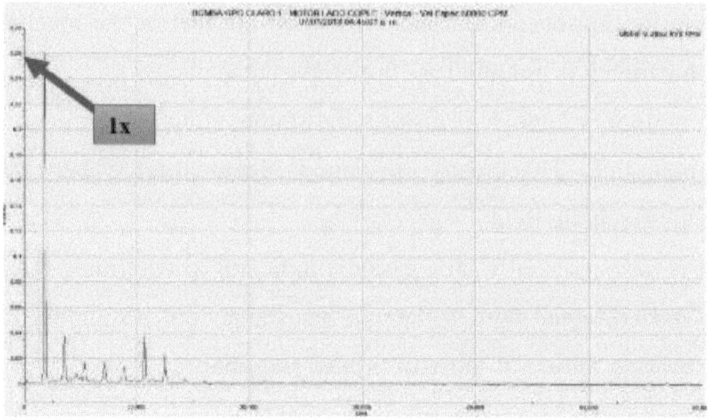

Fig. 3.3.- Vibration spectrum (0.28 in/s), rotational speed of 1,780CPM (1x).

FOLLOW-UP

There are several types of soft or lame feet among the most common are:

1. OSCILLATING (PARALLEL)

Situation where the machine appears to have one leg shorter than the others.

Fig. 3.4.- Lame Foot - Oscillating (parallel)

2. TWISTED (ANGULAR) FOOT

When the machine foot is twisted at an angle in relation to the machine.

Fig. 3.5.- Lame foot - Twisted foot (angular)

3. SOFT (WEAR AND TEAR)

When the machine foot or adjusting plates are already worn and prevent proper seating of the feet.

Fig. 3.6.- Lame foot - Soft (wear and tear)

4. INDUCED

It occurs when an external force such as piping causes the feet of the machine to move.

Fig. 3.7.- Lame - Induced foot

EQUIPMENT FAILURES DUE TO THE LAME FOOT PHENOMENON

The lame foot phenomenon can cause the equipment failures shown in figures 3.8 - 3.10.

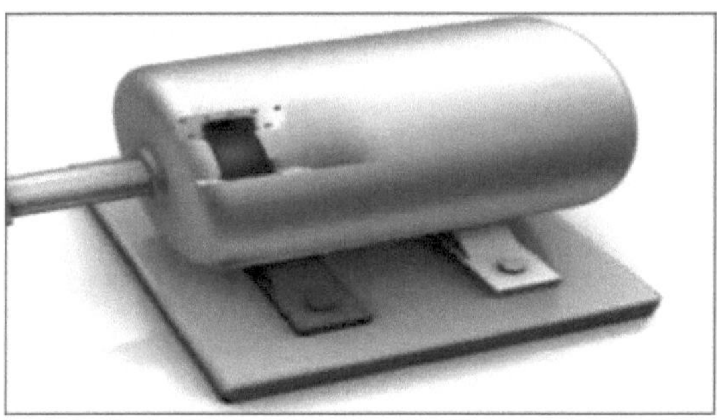

Fig. 3.8.- Misalignment of precision components (the bearings)

Fig. 3.9.- Increase of destructive vibrations, irregular and premature wear of equipment.

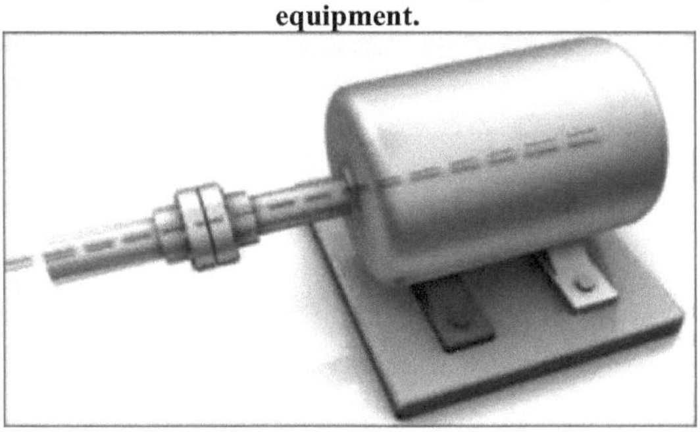

Fig. 3.10.- Deflection of shafts.

FOLLOW-UP

Lame-foot correction procedure.

1. When the engine is found with the four leg bolts loose, the seating of the four legs is checked, then it can be probed with the aid of a shim gauge, if a 0.005" shim fits in any of the legs, it should be compensated with a shim of the same size. After this, all four legs are tightened.

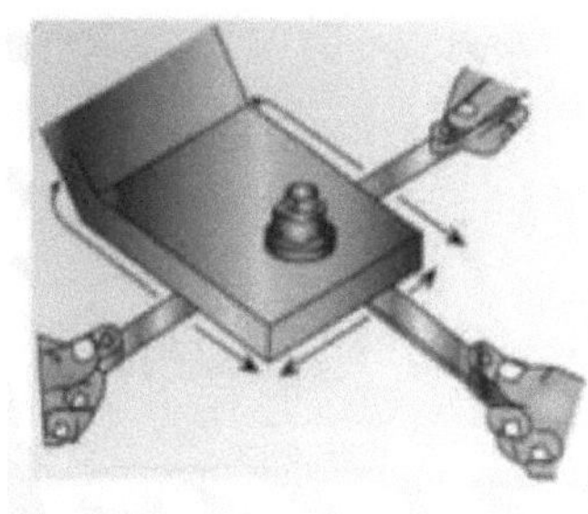

Fig. 3.11.- Checking the seating of the engine feet with shims.

2. Once the four legs are tightened, the caratula indicator is placed on one of the legs, the screw is loosened and the reading of the caratula indicator is taken

when the leg is lifted. If the value is equal to or greater than 0.005", it must be filled with calibrated shims until the clock shows a value of less than 0.005".

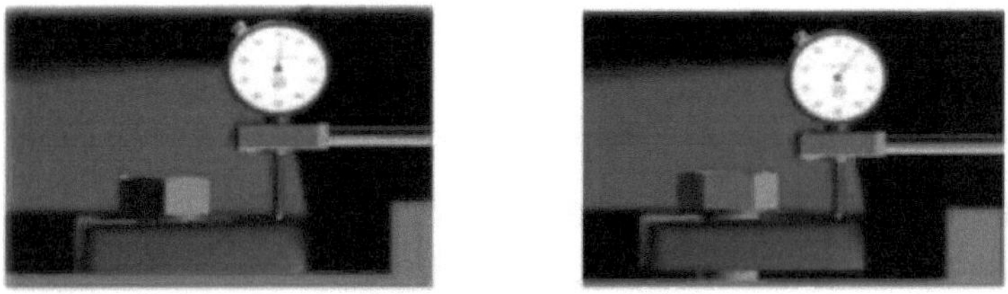

Fig. 3.12.- Taking the reading with the carat meter on the engine feet.

3. Once this action has been corrected, the same procedure is carried out on the following legs 2, 3 and 4.

Fig. 3.13.- Adjusted engine feet.

A very important point is that the shims that were used for the correction of the lame foot must be kept during the alignment.

CASE 4

MISALIGNMENT OF SHAFTS BETWEEN ENGINE ELECTRIC AND SPEED REDUCER (PARALLEL MISALIGNMENT)

The system under test consists of a 70 HP induction motor with a rated speed of 1,775 CPM coupled by an A4 flexible element to a 75 HP planetary speed reducer with a speed ratio of 106.1:1; this transmission moves the No. 4 cane driver through a set of sprockets and chain.

Field data: Regular vibration values in axial and vertical direction with the presence of (2x) i.e. twice the rotational speed of the motor with the highest value in vertical direction. Also the unstable behaviour of the flexible coupling element could be seen with the thermographic camera.

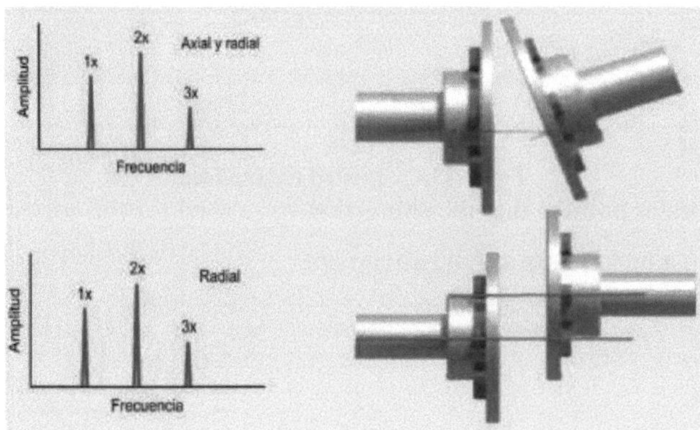

Fig. 4.1.- Misalignment of axles.

The rotation speed is 1,775 CPM (1x), in the spectrum a higher amplitude is perceived at 3,550 CPM, twice the rotation speed (2x), giving by vibration analysis diagnosis a parallel misalignment between the motor shaft and the speed reducer.

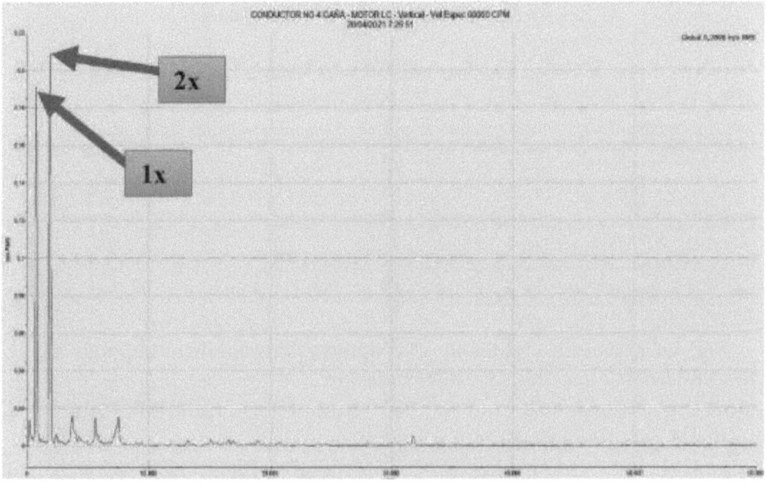

Fig. 4.2.- Vibration spectrum of the 70 HP induction motor.

Once the equipment was stopped, the alignment was checked with the support of a laser axis alignment equipment, and after having carried out the procedure it was observed that:

1. The motor shaft is 0.045" down with respect to the gearbox shaft (out of parameter).
2. The motor shaft is offset 0.005" from the gearbox.

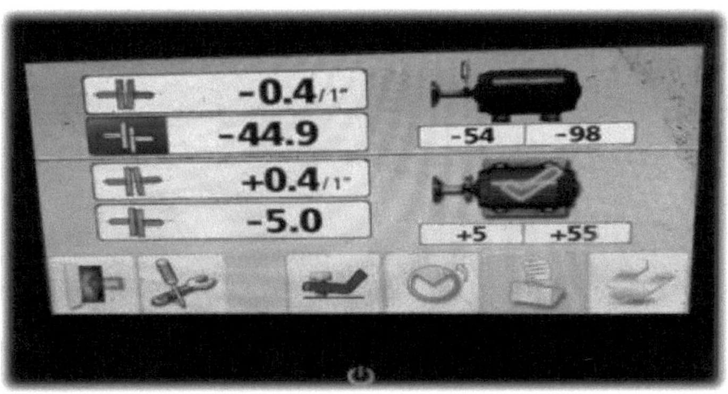

Fig. 4.3.- Reading obtained with the laser alignment equipment.

By consulting the tolerance table on A4 couplings, the permissible misalignment values for this equipment are determined:

DESALINEACIONES ADMISIBLES

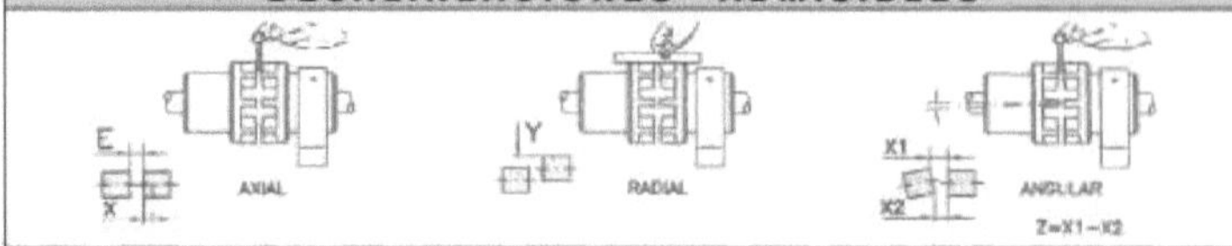

Cotas de montaje (E) y tolerancias en mm

Tipo	A00	A0	A1	A2	A3	A4	A45	A5	A55	A6	A7	A8	A9	A10	A11
E Montaje	1.5	1.5	1.5	2.5	2.5	3.5	3.5	3.5	3.5	3.5	4	5	5	6	6
Axial X	+0.3	+0.3	+0.5	+0.5	+0.7	+0.8	+1.0	+1.0	+1.0	+1.0	+1.0	+1.5	+1.5	+2	+2
Radial Y	0.20	0.30	0.30	0.50	0.50	0.70	0.70	0.70	0.80	0.80	0.80	1.00	1.00	1.20	1.20
Angular Z	0.10	0.10	0.20	0.20	0.30	0.40	0.40	0.50	0.50	0.60	0.90	1.10	1.30	1.70	1.70

Fig. 4.4.- Tolerance tables on A4 couplings Permissible misalignments

From the table it is clear that the tolerances are:

- Radial Y (parallel): 0.70 mm = 0.027" Radial Y (parallel): 0.70 mm = 0.027
- Angular Z (angular): 0.40 mm = 0.015".

Final alignment values of motor shafts and speed reducer shafts

VERTICAL PLANE

- Angular. 0.001"/in.
- Parallel: 0.005" down with respect to the speed reducer shaft.

HORIZONTAL PLANE

- Angular: 0.001"/in.
- Parallel: 0.003" offset from the speed reducer shaft.

After correcting the shaft alignment and replacing the flexible coupling element, the vibration values decreased from 0.28 in/s to 0.05 in/s in amplitude, improving the performance of the equipment.

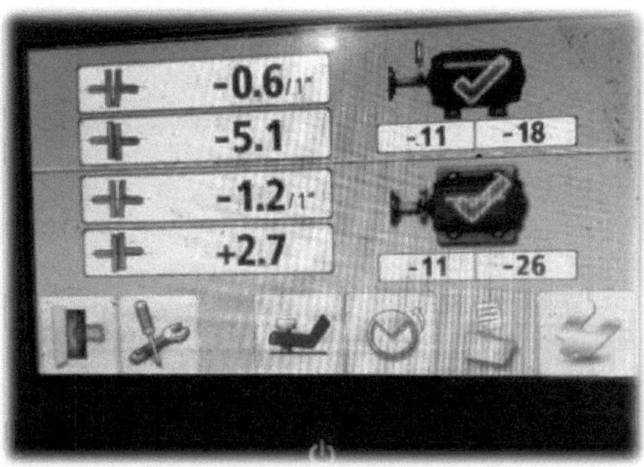

Fig. 4.5.- Reading obtained after making the corrections

Figure 4.6 shows the vibration spectrum after corrections have been made.

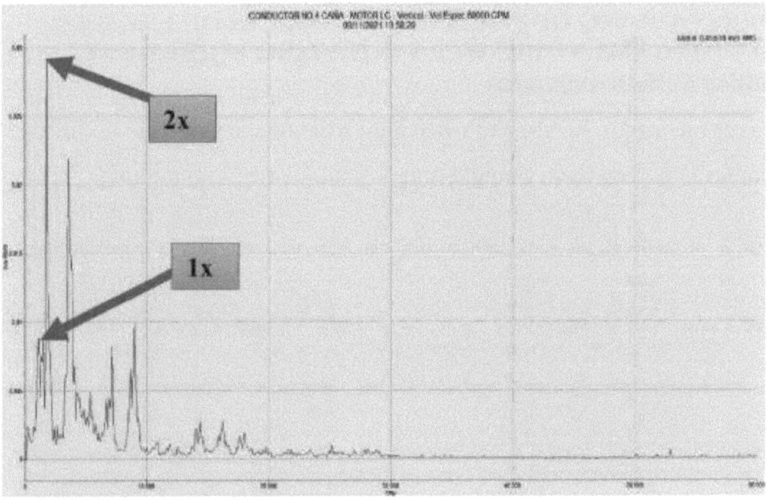

Fig. 4.6.- Vibration spectrum after corrections

FOLLOW-UP

After correcting the shaft alignment and changing the flexible element, it was possible to observe the deterioration of the shaft, and in some cases we can also find other damage caused by the misalignment, among which we can mention:

1. Premature failure of bearing, seal, shaft or coupling.
2. Excessive radial and axial vibration.
3. High housing temperature near the bearings or high discharge oil temperature.
4. Excessive amount of oil leakage at bearing seals or coupling seals.
5. Power loss, motor overheating due to high amperage loads.
6. Heating in couplings or coupling elements.

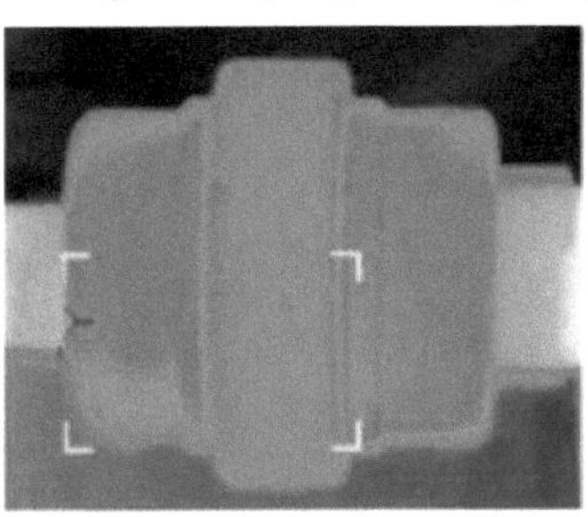

A) Thermographic image of theB) Damage to the flexible element

behaviour of the coupling couple

Fig. 4.7.- Damage caused by misalignment.

UNDERSTANDING OF THE BASIC CONCEPTS OF THE SHAFT ALIGNMENT BETWEEN ROTATING EQUIPMENT

Definition of shaft alignment

Two axes are said to be aligned when their rotational centres are collinear to each other^ i.e. they form a single line.

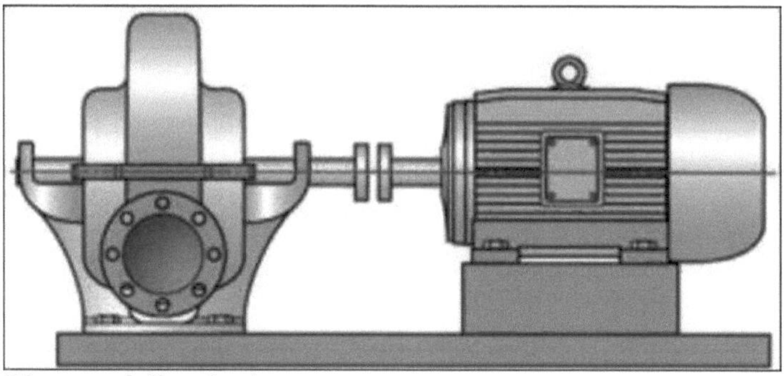

Fig. 4.8.- Diagram of aligned rotary axes.

To carry out the alignment, one of the machines must be taken as fixed (s) and the other as mobile (m), the mobile machine is the equipment that has to be wedged or moved to achieve the collinear line between the centres of the two axes, the fixed or static machine (s) is the equipment that is normally attached to pipes or connections and is complex to move or lift, for this the alignment is carried out in two planes, vertical axis (height) and horizontal axis (displacement).

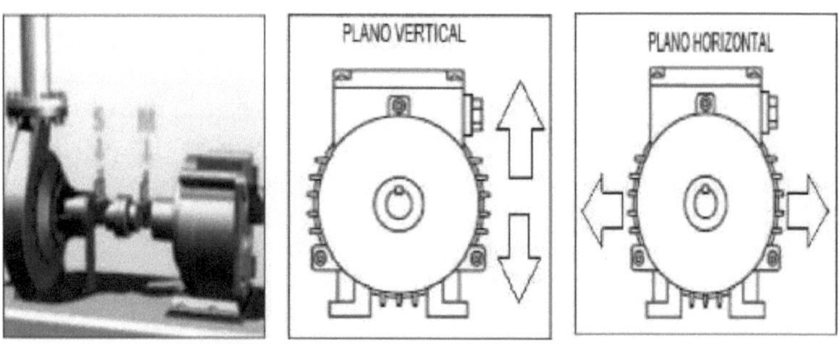

Fig. 4.9.- Alignment on two pianos.

The misalignment can be parallel, angular or combined, i.e. the parallelism as well as the angle of the two axes must be corrected by means of shims placed on the lower part of the legs of the mobile equipment.

To facilitate angular correction it is recommended to correctly level the reference equipment i.e. the fixed equipment(s) in the alignment process, most often pumps or speed reducers.

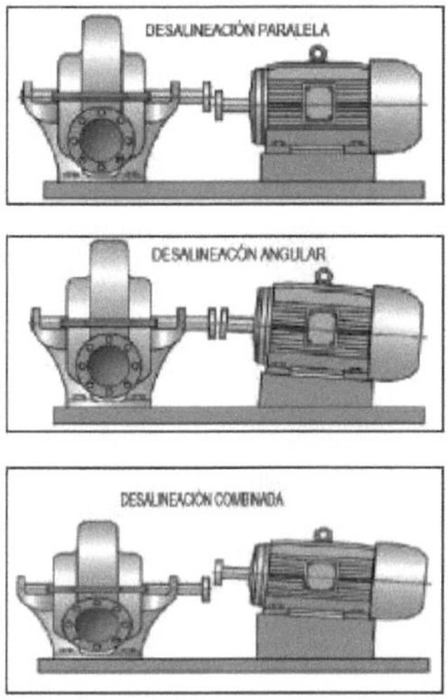

Fig. 4.10.- Types of misalignment.

It is important from the beginning to consider the correction of the soft foot, or lame foot, this consists of checking with a shim gauge that the legs of the motor sit correctly on the base, in case they do not, it is necessary to shim (fill with shims) until the 4 legs of the motor or equipment to be aligned sit correctly (see case 3).

This helps to ensure that when tightening the couplings, the equipment does not move randomly, but maintains a logical trend of vertical and horizontal movements until it comes into alignment tolerance according to the criteria of either coupling tolerance (depending on the type of coupling) or speed.

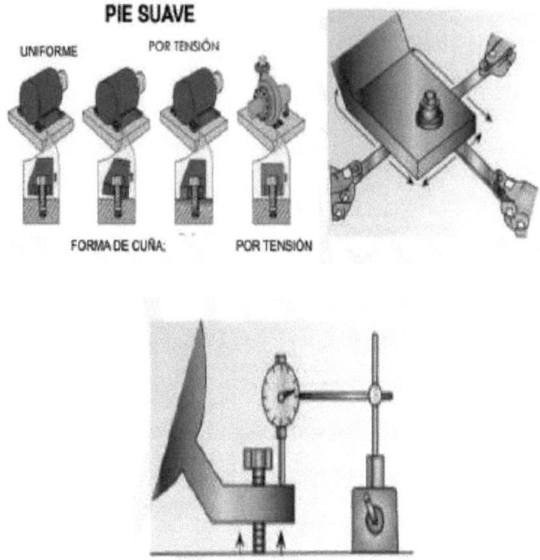

Fig. 4.11.- Soft foot correction (case 3).

CASE 5

MISALIGNMENT OF BEARING 22222 CCK/W33 DUE TO CLEARANCE IN SAF 522 HOUSING (MISALIGNMENT)

The analysed system consists of an induction motor of 250 HP power and a nominal speed of 1,782 CMP that transmits by means of a 13 F coupling, to a cantilever type fan rotor, supported by bearings, coupling side SAF 520 housing with 22220 CCK/W33 bearing and fan side bearing, SAF 522 housing with 22222 CCK/W33 bearing.

Field data: High vibration (0.52 in/s) was felt on fan side bearing in horizontal direction, taking a temperature of 82 °C.

Recently the equipment had been uncovered for lubrication and the top cover was found to be out of adjustment and a 0.002" was fitted to reduce the fit of the housing, the 22220 CCK/W33 bearing was replaced due to damage to the cage, to remove the bearing the electric motor was removed from the base.

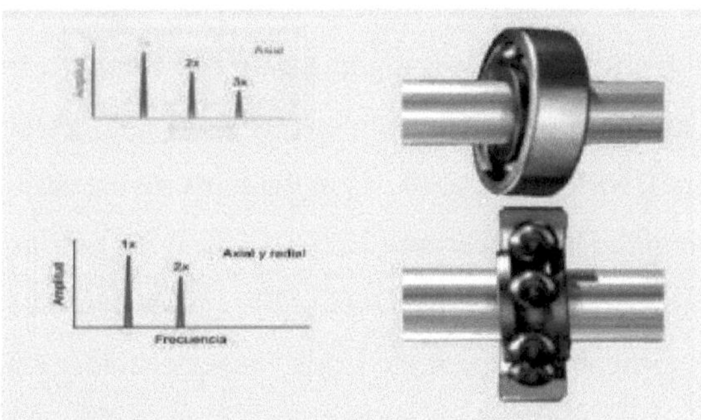

Fig. 5.1.-Vibration in bearing 22220 CCK/W33

A high amplitude value (0.54 in/s) was observed in the vibration spectrum in the horizontal direction, frequencies of 1,783 CPM (1x), 2,766 CPM (2x) and 5,349 CPM (3x) could be seen in the spectrum. By vibration analysis it was determined as misalignment of the bearing in the housing due to looseness or misalignment in the housing (SAF 522 housing).

It was recommended to measure with plastigage the total misalignment of the upper cover of the SAF 522 housing against the outer bearing race, as well as the alignment between the two bearings.

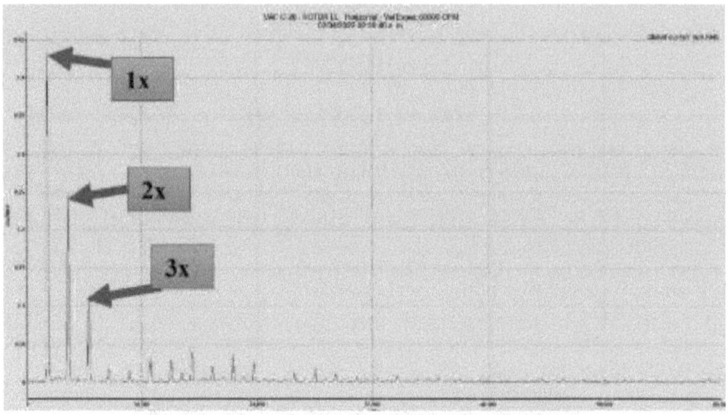

Fig. 5.2.- Espectro de vibración en el rodamiento

Symptoms in a vibration spectrum for bearing misalignment are the occurrence of (1x), (2x) and (3x) in the radial direction. This is a unique form of looseness caused by loads on the bearing pedestals, loose pedestal pins or misaligned or defective bearing housings.

The vibrations will have a strong (1x) and (2x) component due to the movement generated by the machine and the peak (2x) may be higher than the peak (1x) which may lead to suspicion of misalignment. In some cases a frequency (3x) is generated. The spectrum will have components at (1x), (2x) and (3x) (but often no more harmonic), with a peak at (0.5x) in the most severe cases.

Previously, a coupling side bearing (22220 CCK/W33 bearing) was replaced due to damage to the cage, some of the possible causes could be poor mounting (misalignment of the bearing), high loads, excessive rotational speed, poor lubrication and increased temperature.

Fig. 5.3.- Bearings supporting the rotor of the electric motor.

With the support of the thermographic camera, it was possible to observe temperature values of 82 °C in the fan side bearing, as well as temperature of 81 °C in the bearing on the coupling side of the motor. It is important to check the alignment point by point, starting with the alignment between the bearing on the coupling side and the fan side, as well as the alignment between the fan shaft and the motor shaft.

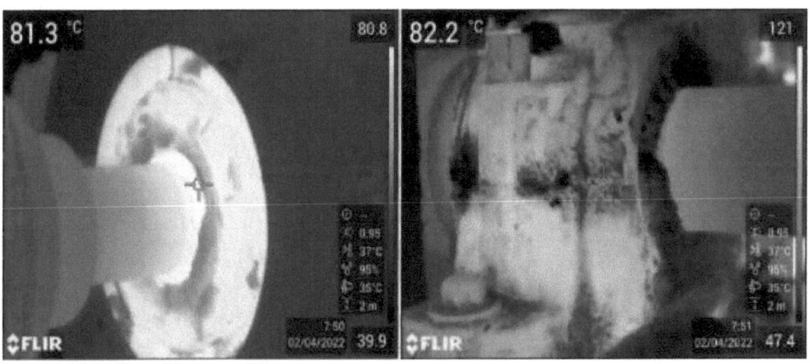

A) Fan sideB) Coupling side

Fig. 5.4.- Thermographic image of the bearings of the electric motor.

When uncovering the bearing housing on the fan side, it was possible to observe bearings with marks on the outer race, as well as rust, derived from the misalignment in the housing (SAF 522 box), we proceeded to measure the adjustment with the help of plastigage, marking a value of 0.23 mm (0.009"), so it was considered to shim the top cover with a 0.008" shim to compensate the misadjustment.

Fig. 5.5.- Fan side bearing with marks on outer race.

When limiting the clearance of the upper housing, it was necessary to verify the reduction of the radial clearance of the bearing, for which the data pertaining to the 22222 CCK/W33 bearing provided by the manufacturer was taken.

The locking nut was tightened by forcing the bearing up the sleeve until the measured bearing clearance was reduced to the values indicated in the table shown in figure 5.6.

Size* Size* Size* Size* Size* Size* Size* Size* Size* Size* Size* Size* Size*	IREDUCTION OF	BEARING	SPACE1	
09	0.0010	0.0012	0,025	0,030
10-13	0.0012	0.0016	0,030	0,040
15-16	0.0016	0.0020	0,040	0,050
17-20	0.0018	0.0024	0,045	0,060
22-24	0.0020	0.0028	0,050	0,070
26-28	0.0026	0.0035	0,065	0,090
30-32	0.0030	0.0039	0,075	0,100
34-36	0.0031	0.0043	0,080	0,110
38-40	0.0035	0.0051	0,090	0,130
44	0.0039	0.0055	0,100	0,140
48	0.0043	0.0059	0,110	0,150
52-56	0.0047	0.0067	0,120	0,170
60	0.0051	0.0075	0,130	0,190
64-68	0.0059	0.0083	0,150	0,210
72-80	0.0067	0.0091	0,170	0,230

*The size is represented by the last two bearing numbers of the bearing being installed.

Fig. 5.6.- Bearing clearance values suggested by the manufacturer

According to the values in the table shown in figure 5.6, we have:

- Bearing 22222 CCK/W33 unmounted space = 0.0070".
- Bearing clearance reduction = 0.0020" TO 0.0028".
- Maximum bearing radial clearance = 0.0070"-0.0020" = **0.0050".**
- Minimum bearing radial clearance = 0.0070" -0.0028" = **0.0042".**

Locating bearings are placed with the bearing face opposite the lock nut against the flange of the bearing housing.

Normally, expansion bearings are centred in the bearing seal, between the flanges, to allow the shaft to expand.

Shafts with more than one bearing have only one fixed bearing, so the fixed unit is usually installed next to the transmission. The fixed bearing absorbs all thrust loads.

The bearing seal grooves in the bearing cover are greased and the bearing is placed on top, after cleaning the contact surfaces.

The two pins align the cover with the lower half of the bearing housing. Each of the caps must match its lower half, as these parts are not interchangeable.

Tighten the bolts completely with a torque spanner, according to the values given in the table in figure 5.7; apply oil only to the bolt threads.

Tightening torque of bolts without bearing nut

Bearings	Ta ma no pern o	Torque (ft-lb)	Torque (N m)
SAF509-510	MIO	30	40
SAF511-513	M12	60	51
SAF515	M12	60	BI
SAF516	M12	60	81
SAF517	M16	125	170
SAF513	M12	60	BI
SAF52O - 522	M16	125	170
SAF524	M16	125	170
SAF526-528	M2O	250	339
SAF53O	M2O	250	339
SAF532	M2O	250	339
SAF534	M24	500	678
SAF536-538	IM 24	500	678
SAF54O	M24	500	678
SAF54 4	M24	500	678
SAFO43	M24	500	678
SAFO52	IM 24	500	678
SAFQ56	M3O	750	1,017
SDAF152	M3O	750	1,017
SDAF156 - 160	M3O	750	1,017
SDAF164	IM 42	2,000	2,712
SDAF168 - ISO	M42	2,000	2,712
SDAF-1B4	M42	2,000	2,712
SDAF 232/500	M48	2,860	3,878

Fig. 5.7.- Tightening torque of bolts without bearing nut suggested by the manufacturer.

With the mounting bolts not yet fully tightened, alignment and ease of rotation were checked.

By making a visual inspection of the space between the shaft and the seal bore, the distance between the outside diameter of the seal and the diameter of the enlarged bearing bore was measured at three points. Each measurement should be taken at 90 degrees to the previous measurement.

All three measurements must be uniform to ensure that the alignment is correct; alignment was done using cradles or by changing parts as necessary.

It is recommended to use stops against the faces or ends of the bases located opposite to the direction of load or vibration to prevent bearing movement.

Finally the mounting bolts are tightened firmly. To install cast iron bearings, UCN mounting bolts up to grade 5 SAE can be used with appropriate torque.

The alignment between the bearings on the fan side and the coupling side was checked, taking as a reference point the central point of the upper cover of the coupling side bearing and projecting the laser through the centre of the date until it reached the other bearing.

Alignment correction was performed by running the bearing on the coupling side 1/8" to match the measurement between the bearing marks on each side (22.375").

Fig. 5.8.- Alignment of bearing centres with the support of laser alignment equipment.

After aligning the bearings, the alignment between the fan shaft and the electric motor shaft was checked and corrected with the help of the laser shaft aligner, the acceptable alignment values are taken from the speed table (1,800 RPM), since the speed of the electric motor of this fan is 1,782 CPM.

DIMENSIONES (MILÉSIMAS DE PULGADA)				
RPM	DESALINEAMIENTO ANGULAR		DESALINEAMIENTO PARALELO	
	EXCELENTE	ACEPTABLE	EXCELENTE	ACEPTABLE
< 900	1.0	1.5	4.0	8.0
1200	0.7	1.0	3.0	6.0
1800	0.5	0.7	2.0	4.0
> 3600	0.3	0.5	1.0	2.0

	MISALIGNMENT ANO OLAH		DEFROSTING rARAL ELO	
	EXCELLENT	ACCEPTABLE	EXCELLENT	ACCEPTABLE
< 900	-1.0	-1.5	4.0	8.0
*1200	0.7	-1.0	3.0	6.0
*1800	0.5	0.7	2.0	4.0
> 3600	0.3	o.s	*1.0	2.0

Fig. 5.9.- Acceptable alignment values recommended by the manufacturer.

After the readjustment of the 22222 CCK/W33 bearing housing, measurements were taken.

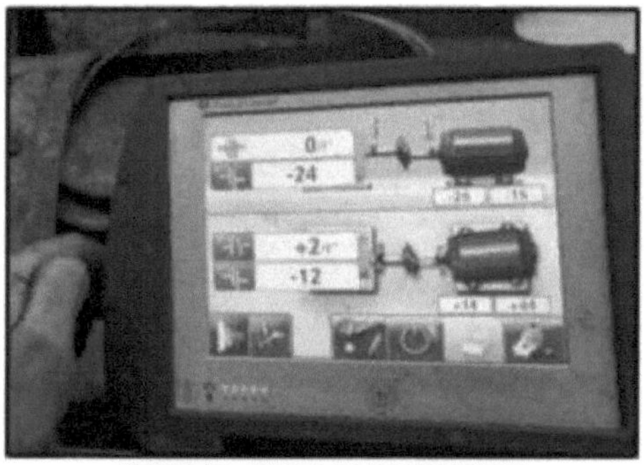

INITIAL MEASUREMENT

HORIZONTAL PLANE

VERTICAL PLANE

Fig. 5.10.- Alignment measurements obtained before and after the realignment work after the readjustment works

The values obtained are shown in figure 5.9.

FINAL VALUES OF MOTOR AND FAN SHAFT ALIGNMENT.

VERTICAL PLANE

ANGULAR. **0.000 "IPLG**

PARALLEL: **0.000"** BELOW THE SPEED REDUCER SHAFT.

HORIZONTAL PLANE

ANGULAR: **0.000"/P LG.**

PARALLEL: **0.003"** OFFSET FROM THE SPEED REDUCER SHAFT.

Fig. 5.11.- Final alignment values of motor and fan shafts.

Figure 5.12 shows the vibration spectrum after readjusting the 22222 CCK/W33 bearing housing, checking the alignment between bearings and aligning the fan rotor and electric motor shafts, achieving a decrease in the overall vibration amplitude (from 0.54 in/s to 0.25 in/s).

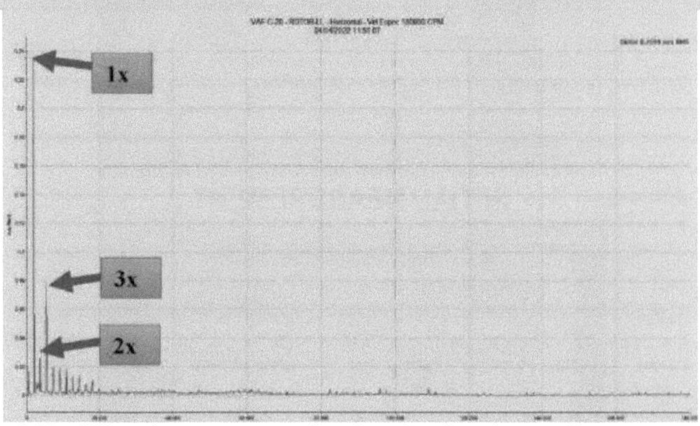

Fig. 5.12.- Vibration spectrum after bearing housing realignment work.

CASE 6

LOOSENESS ON ELECTRIC MOTOR BASE (MECHANICAL CLEARANCE TYPE "A").

The system under consideration consists of a 50 HP induction motor with a rated speed of 1,780 CPM coupled through a 10 F coupling to a 50 HP, 800 GPM, closed impeller centrifugal pump, which pumps water from a compact cooling tower to the power plant.

Field data: High amplitude motor vibration (0.79 in/s) at motor rotational speed 1,780 CPM (1x) in horizontal direction, values out of range, newly installed equipment consists of a base plate which is fixed by anchors and grouting, the motor base is welded to this base plate.

Fig. 6.1.- Closed impeller centrifugal pump of 50 HP power and 800 GPM.

The vibration spectrum shows a high amplitude value (0.76 in/s) at a rotational speed of 1,780 CPM (1x), which could be assumed to be an unbalance, but caused by the looseness of the base plate. On closer inspection, points were identified where the motor base was not attached to the base plate.

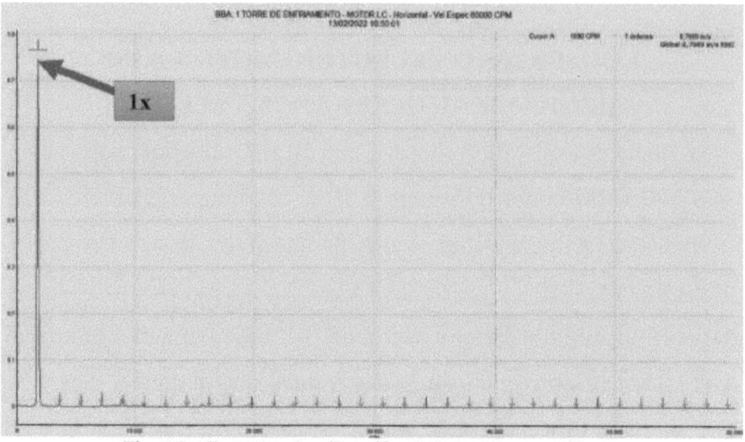

Fig. 6.2.- Espectro de vibración de la bomba centrifuga.

Amplitude of 0.76 in/s @ 1,780 CPM

After inspecting the base it was observed that it was not welded correctly, between the base of the motor and the base plate there are gaps without welding causing vibration in the motor.

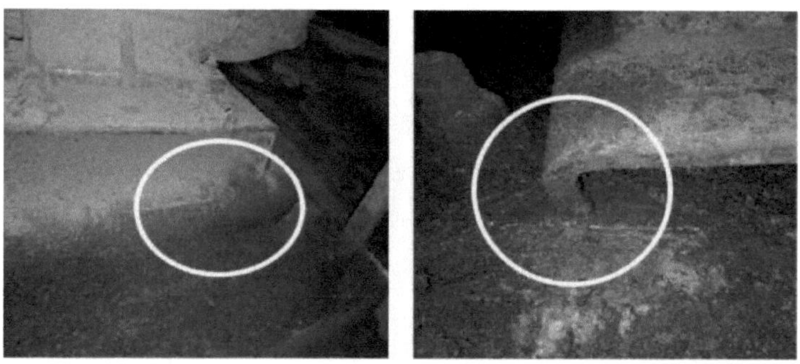

Fig. 6.3.- Motor base loose from the base plate (lack of welding).

FOLLOW-UP

Type "a" mechanical clearance characteristics (mechanical or structural clearance).

These are the clearances associated with the non-rotating mechanical parts of the machine: anchorages for fixing to the bedplate, joints between pipes, bearing housings, etc.; they usually occur more in the radial directions of measurement than in the axial directions, with the presence in the frequency spectrum of several harmonics of the rotational speed depending on the severity of the clearance.

Phase analysis can reveal approximately 90° to 180° of phase difference between vertical measurements on the bolt, machine stand or the base itself.

The rotor represents an unbalance which acts as an excitation force of the clearances due to the loosening of the anchorages between the bearing support and the bedplate, the unbalance observed in the spectrum (1x) is caused by the instability of the rotor to exert its rotation on its centre.

In practice the attempt to swing a rotor that is loose (mechanical looseness) causes the phase angle and the weight value calculated by the vibration equipment to change and therefore not respond to the swing, being a consequence of another cause in this case the looseness of the base plate.

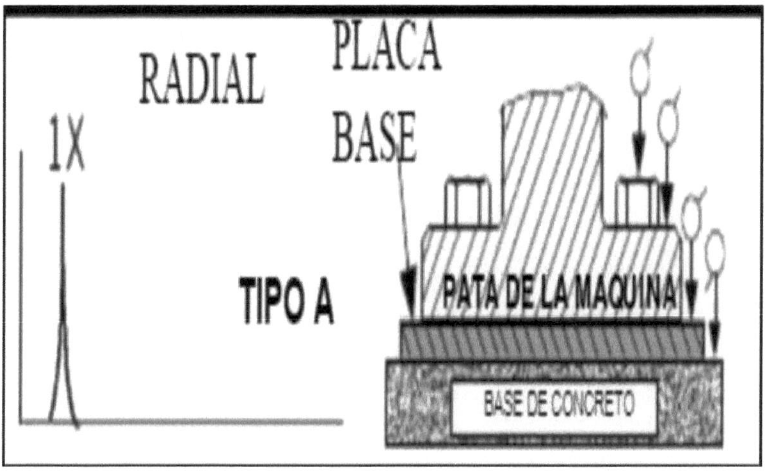

Fig. 6.4.- Fixing of an engine to prevent mechanical loosening.

It was recommended to weld on clamps or angle brackets to reduce vibration and give stability to the engine in operation.

The mechanical looseness causes a lame-foot or wobble effect which amplifies the vibration at the rotational speed (1,780 CPM).

Fig. 6.5.- Welding clamps added to the engine.

The overall vibration values decreased, after fixing the base plate to the motor base, the vibration amplitude decreased from 0.76 in/s to 0.06 in/s, i.e. by up to 700%.

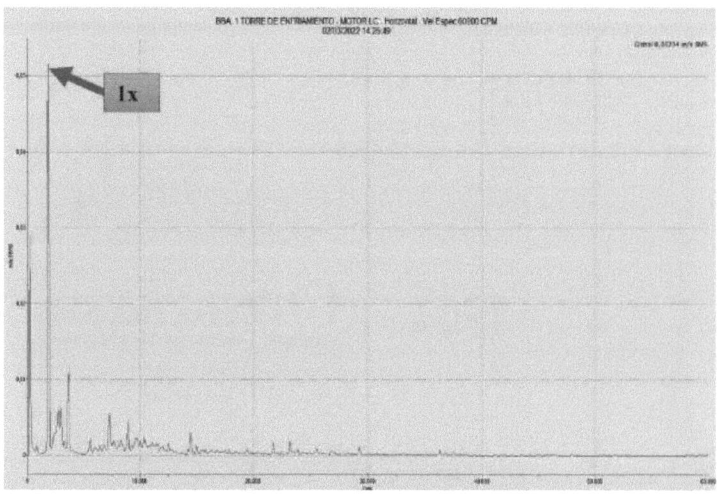

Fig. 6.6.- Vibration spectrum of the centrifugal pump after corrections. Amplitude of 0.06 in/s @ 1,780 CPM.

A

ACCELERATION

Ratio of change of velocity to time.

ALIGNMENT

Position in which the centre lines of two axes should be as collinear as possible when the machine is operated.

AMPLITUDE

The maximum value of a sine wave.

SPECTRAL ANALYSIS

It is the interpretation of a spectrum to determine the physical meaning of what is happening in a machine.

ARMONICO

These are vibrational frequencies that are multiple integrals of a specific fundamental frequency.

FRACTIONAL HARMONIC

Harmonics that lie between the main harmonics and are fractions of the fundamental frequency.

AXIAL

Position of the sensor going in the direction of the shaft line.

B

BALANCING

A process by means of which the centre of mass of a rotor is made to coincide with the centre of rotation.

CYCLE

It is a range of values over which a periodic phenomenon repeats itself.

D

SCROLLING

Change of position of an object or particle according to a reference system.

DIAGNOSIS

The process by which the condition of a machine is judged.

FREQUENCY DOMAIN

It is the graphical representation of a vibration signal in which amplitude vs. time is plotted.

ENTREHIERRO

Air space between the stator and the rotor of an electric motor.

SPECTRUM

Synonymous with frequency domain.

EXCENTRICITY

Variation of the centre of rotation of the shaft with respect to the geometrical centre of the rotor.

F

SERVICE FACTOR

Current factor that corrects for standardised levels, for special machines that require particular levels.

PHASE

It is a time delay of two signals, expressed in degree of rotation.

FREQUENCY

It is the ratio of the period and means the number of complete oscillations per unit time.

CAGE FAILURE FREQUENCY (CFF)

The frequency of a bearing that is excited when it has damage to its cage.

ROLLING ELEMENT FAILURE FREQUENCY (BFS)

The frequency of a bearing that is excited when damage occurs to a rolling element.

INTERNAL TRACK FAULT FREQUENCY (BPFI)

The frequency of a bearing that is excited when damage to the inner race occurs.

EXTERNAL TRACK FAULT FREQUENCY (BPFO)

The frequency of a bearing that is excited when damage to the outer race occurs.

NATURAL FREQUENCY

It is the frequency that each component represents by its very nature. This frequency will oscillate if excited by an external agent operating at a frequency very close to it.

H

HORIZONTAL

Generally, it is the position given to the sensor, which is perpendicular to the direction of gravity.

MICRA

A measure of length or distance. equivalent to one thousandth of a millimetre.

MILS

A measure of length or distance equivalent to one thousandth of an inch.

P

PERIOD

The time required for an oscillation to occur or a cycle to complete. Generally, it is given in minutes and seconds.

PICO

Each of the spectrum's constituent lines.

RADIAL

Position of the sensor running perpendicular to the axis line.

RESONANCE

It occurs when the natural frequency of a component is excited by an external agent: the vibration amplitude of the machine can increase enormously, causing damage that can be catastrophic at some point.

FLEXIBLE ROTOR

They are rotors that rotate very close to their critical speed and therefore show significant deformation.

R^GID ROTOR

Rotor that does not significantly deform when operating at its rated speed.

RPM
Another unit of frequency is the number of cycles per minute of the machine.

NOISE

It is information of the signal that does not represent some importance, represented as signal contamination.

SUBARMONIC

These are frequencies that are within a fixed fraction of a fundamental frequency, such as the rated speed of the machine.

SPEED
Ratio of change in displacement to time.

NOMINAL SPEED
Input speed of a machine.

VERTICAL
Position given to the sensor, in the sense of the acceleration of gravity.

VIBRATION
It is an oscillatory movement.

RANDOM VIBRATION
Frequency that do not comply with special repeating patterns.

TABLE: VIBRATION SEVERITY RANGES ACCORDING TO ISO 10816.

CASE 1.- UNBALANCE IN HORIZONTAL SWING-BACK TYPE BLADE-HOLDING ROTOR (UNBALANCE).

- Development Of A Balancing System For Low Power Turbines, Ing. Ruben Eduardo Sosa.
- Predictive 21. Digital Journal of Maintenance, Reliability and Asset Management.

CASE 2.- STEAM TURBINE ROTOR UNBALANCE (UNBALANCE AND MECHANICAL LOOSENESS).

- REXNOR, Installation and Maintenance Instructions, Falk Steel Flex®.
- POWER-MI Vibration Analysis Manual.
- Mobius Institute ™, Mobius Ivibe.

CASE 3.- SOFT OR LAME FOOT ON ELECTRIC MOTOR (MECHANICAL BACKLASH TYPE "A").

- CBM Connect, Bill Kruger, AllTest Pro Training Technical Manager.
- PROFTECHNIK /What Is The Lame Foot Phenomenon?
- POWER-MI Vibration Analysis Manual.

CASE 4.- MISALIGNMENT OF SHAFTS BETWEEN ELECTRIC MOTOR AND SPEED REDUCER (PARALLEL MISALIGNMENT).

- REXNOR, Installation and Maintenance Instructions, FALK STEELFLEX®.
- POWER-MI Vibration Analysis Manual.
- Images, MYG Inc.

CASE 5.- MISALIGNMENT OF BEARING 22222 CCK/W33 DUE TO PLAY IN THE SAF 522 HOUSING (MISALIGNMENT).

- PPI, Precision , Puller & Idler, Operation & Maintenance Bearings & Bearings SAF.
- REXNOR, Installation and Maintenance Instructions, Falk Steel Flex®.
- Capacitacion Mecanica Industrial, Ing. Gonzalo Morales Melendez.
- POWER-MI Vibration Analysis Manual.
- Mobius Institute ™,Mobius Ivibe.

CASE 6.- ELECTRIC MOTOR BASE CLEARANCE (MECHANICAL CLEARANCE TYPE "A").

- Technical Associates of Charlotte, Pc. R-0894-4.
- POWER-MI Vibration Analysis Manual.

MIX
Papier aus verantwortungsvollen Quellen
Paper from responsible sources
FSC® C105338

Printed by Books on Demand GmbH, Norderstedt / Germany